POLLUTION

ANNE O'DALY

BROWN BEAR BOOKS

Published by Brown Bear Books Ltd
4877 N. Circulo Bujia
Tucson, AZ 85718
USA

and

Studio G14, Regent Studios,
1 Thane Villas, London N7 7PH, UK

ISBN 978-1-78121-817-4 (library bound)
ISBN 978-1-78121-823-5 (paperback)

Library of Congress Cataloging-in-Publication Data available on request

Design: squareandcircus.co.uk
Design Manager: Keith Davis
Children's Publisher: Anne O'Daly

Manufactured in the United States of America
CPSIA compliance information: Batch#AG/5652

Picture Credits

The photographs in this book are used by permission and through the courtesy of:

iStock: Alex Punter 16–17; Shutterstock: Aevan-Stock 18–19, Anticiclo 8–9, Nancy Ginzburg 4–5, okanozdemir 6–7, ohrim 12–13, overcrew 10–11, vchal 14–15, WAYHOME Studio 20–21.

t-top, r-right, l-left, c-center, b-bottom

Brown Bear Books has made every attempt to contact the copyright holder. If you have any information about omissions please contact: licensing@brownbearbooks.co.uk

Websites

Words in **bold** appear in the Glossary on page 23.

CONTENTS

WHAT IS POLLUTION?

Pollution makes our planet dirty. Harmful substances get into the environment. They are called pollutants. They make pollution. It affects our air, water, and land.

WHAT MAKES POLLUTION

Some pollution comes from natural things. Volcanoes send ashes into the air. Forest fires make smoke. But most pollution is made by human activities. Burning **fossil fuels** makes air dirty. Chemicals from factories run into rivers. The trash we throw away pollutes land.

Cars run on fossil fuels. They release harmful gases. This pollutes the air.

Types of Pollution

Pollution comes in different forms. Here are the main ones.

Air pollution

Water pollution

Land pollution

Noise pollution

Light pollution

AIR POLLUTION

Most air pollution comes from fossil fuels. We burn them for power. Burning fossil fuels sends harmful gases into the air. One of the gases is **carbon dioxide**. It goes into the **atmosphere**. It traps the Sun's heat. This makes the planet heat up.

SMOG

Other gases make smog. This is a dark cloud. It hangs over cities. Smog can make it hard to see. It can make people cough. It can make it hard to breathe.

Smog forms on hot days. It's looks like a thick, brown haze. Smog harms plants and people.

Ozone

Ozone is a gas. It is made of oxygen. A layer of ozone is high in the atmosphere. This ozone layer protects us from the Sun. But ozone near the ground makes smog. Ozone pollution is worse on sunny days.

How Smog Forms

Gases from cars and factories go into the air. They react with sunlight and ozone to make smog.

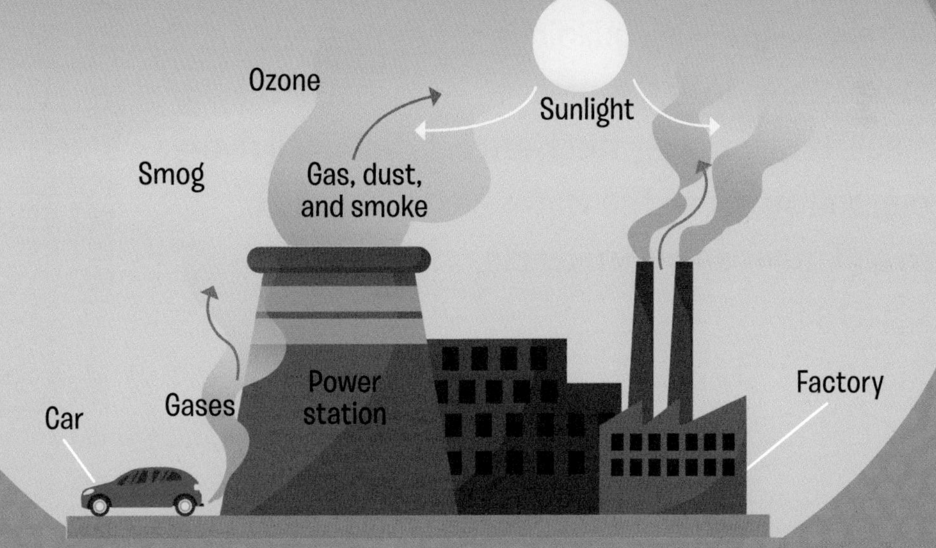

ACID RAIN

Air pollution also makes acid rain. Harmful gases rise in the atmosphere. They mix with air and water. They turn into acids. The acids fall to the ground in rain and snow.

BUILDINGS AND TREES

Acid rain wears away buildings. It harms trees. It makes them lose their leaves. Acid rain damages the soil. It takes away **nutrients**. It falls into rivers, lakes, and streams. The water becomes acidic. Fish and other animals may die.

Dying trees are a sign of acid rain. The rain takes nutrients from the soil. Trees need nutrients to grow.

Acid Rain

Acid rain forms from air pollution. It kills trees and poisons water.

Harmful gases are released into the air.

The gases mix with water and make acid.

Acid rain kills trees and goes into the soil.

Acid in water kills fish.

RIVERS AND STREAMS

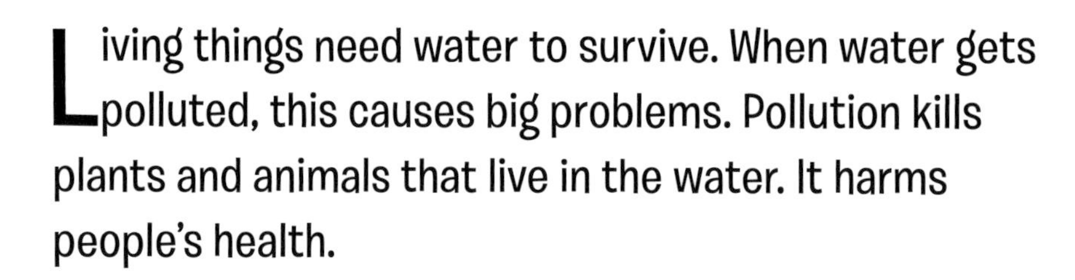

Living things need water to survive. When water gets polluted, this causes big problems. Pollution kills plants and animals that live in the water. It harms people's health.

HARMFUL CHEMICALS

Chemicals from factories run into rivers. **Pesticides** and **fertilizers** are used on farms. Some get washed into streams. Household cleaners go down the drain. Another cause of pollution is **sewage**. It gets into rivers and streams. Sewage contains harmful **bacteria**. People who drink the water may get sick.

People dump garbage in rivers and streams.

What Makes River Pollution

acid rain

sewage

farmyard waste

chemicals from farms

waste from factories

POLLUTING THE OCEANS

Oceans cover 70 percent of Earth. They are home to all kinds of plant and animal life. But our oceans are getting polluted.

COMING FROM LAND

Most ocean pollution comes from land. It runs into the sea from rivers. Some is dumped straight into the ocean. Oil tankers can spill oil. The oil floats on the water. It poisons fish and other sea creatures. Birds get coated with oil. The oil damages their feathers. The birds can't fly or dive for food.

Spilled oil ends up on beaches. Cleaning up is hard work.

Pollution on a Plate

Oil in the ocean can even affect people.

Oil gets into tiny plants and animals

Small fish eat the tiny creatures

Big fish eat the small fish

People eat the big fish

POLLUTING THE LAND

Landfill sites are filling up. We are running out of space for new ones.

Land gets polluted when trash gets dumped in the ground. Some of our garbage is **recycled**. But most is buried in **landfill** sites. Chemicals from the trash can leak into the land. Some pollutants break down. Others stay in the soil forever.

POLLUTION FROM FARMS

We need farms for food. But farms can cause land pollution. Farmers spray pesticides and fertilizers on crops. The chemicals run into the soil. They get washed into water deep under the ground.

How Long?

Some items break down quickly in landfill. Others can take hundreds of years.

Banana peel
2 to 5 weeks

Paper towel
2 to 4 weeks

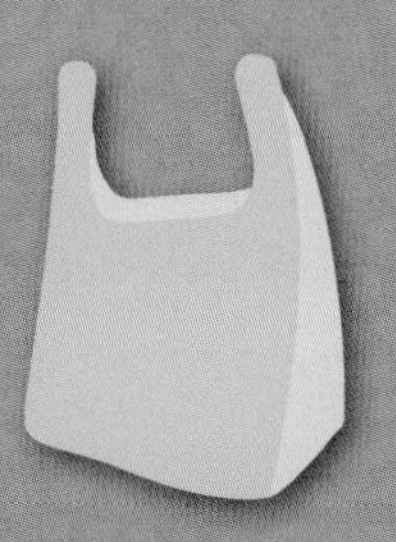

Plastic bag
10 to 20 years

Cotton t shirt
2 to 5 months

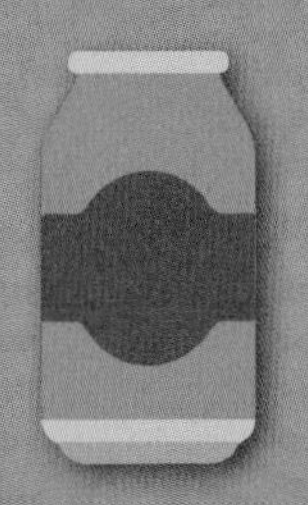

Aluminum can
80 to 200 years

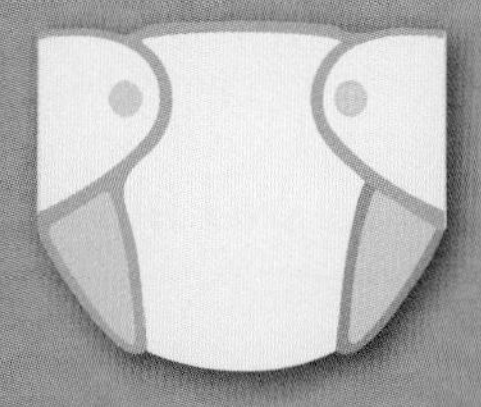

Disposable diaper
450 years

PLASTIC PROBLEM

Plastic is everywhere. It can be made into lots of things. But plastic doesn't break down. Almost all the plastic ever made is still on Earth.

PLASTIC WASTE

Plastic waste is a big problem. Most ends up in landfill. It pollutes the land. Plastic also gets into the oceans. It kills sea animals. They get tangled in plastic bags. Some animals think the plastic is food. They eat it. The plastic fills their stomach. The animals think they are full. They stop eating and starve.

Sea turtles feed on jellyfish. They eat plastic bags thinking they are jellyfish.

Say No to Single-Use

A lot of plastic trash has been used just once. Here are some ways to stop using single-use plastic.

Swap plastic sandwich bags for a reusable lunchbox

Take a cloth bag to the store

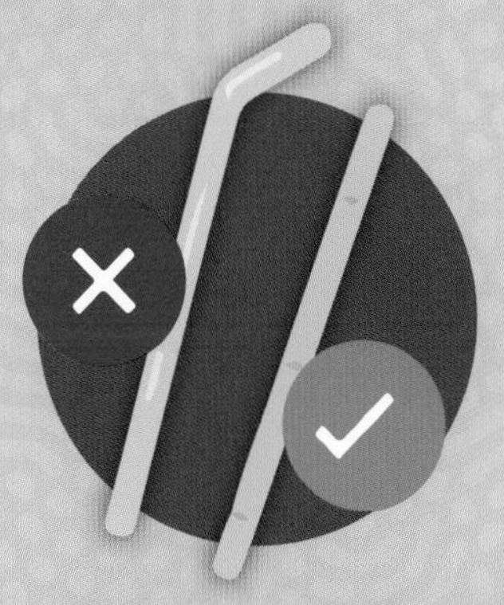

Ditch plastic straws for paper ones

Use a refillable water bottle

NOISE AND LIGHT POLLUTION

Noise and light also make pollution. Loud noise makes noise pollution. Traffic makes a lot of noise. So does building work. Noise at night stops people sleeping. Loud noise can damage people's hearing.

Bright lights and billboards light up a city at night. Too much light is a kind of pollution.

BRIGHT LIGHTS

Too much light makes pollution. Most comes from street lights and bright advertisements. Light pollution stops people sleeping. Bright lights make it hard to see the night sky. Some birds and animals use the moon to find their way. Light pollution makes them get lost.

Lower the Lights

We can cut down light pollution. Only turn on outdoor lights when they are needed. Make sure they point down rather than up into the sky. At night, close curtains and window blinds to keep light in.

How Loud?

Noise is measured in decibels. This scale shows the level of some common sounds.

0-40db Faint	30db	**Leaves rustling**
50-60db Moderate	30db	**Conversation**
70db Loud	70db	**Car**
80-100db Very loud	90db	**Hair dryer**
110-140db Extremely loud	140db	**Fireworks**

WHAT CAN WE DO?

We can all do something to reduce pollution. Cars send harmful gases into the air. Cycle, walk, or scoot to school. Use public transportation instead of cars.

RECYCLE

A lot of our trash can be recycled. Recycling saves energy. It stops trash ending up in landfill. It helps stop pollution. Recycle as much as possible. Buy recycled goods when you can.

Join a beach clean-up. Clean-ups are fun! And they make beaches safer for people and wildlife.

What Can Be Recycled

Check this list to see what you can recycle.

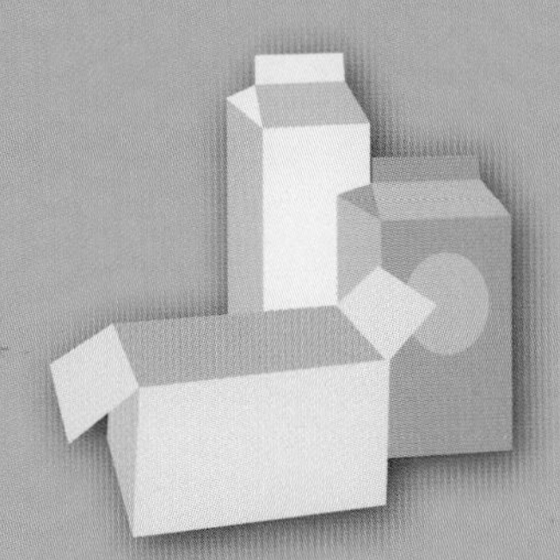

Cardboard boxes and cartons

Paper and newspaper

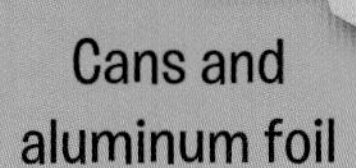

Cans and aluminum foil

Glass bottles and jars

Some plastic bottles and containers

QUIZ

How much have you learned about pollution? It's time to test your knowledge!

1. What is smog?

a. a thick, dark cloud that hangs over large cities

b. a type of fossil fuel

c. oil that has spilled into water

2. What makes acid rain?

a. a mix of plastic and water

b. chemicals that run from farmland into rivers

c. harmful gases that dissolve in water high in the atmosphere

3. How much of the pollution in the ocean comes from land?

a. One-tenth

b. More than four-fifths

c. None of it

4. What does the decibel scale show?

a. The size of large bells

b. The amount of light that comes from a light bulb

c. How loud noises are

The answers are on page 24.

GLOSSARY

atmosphere the blanket of gases that surrounds Earth

bacteria life-forms that are too small to see by eye

carbon dioxide (CO_2) a gas that humans and animals breathe out, which is also produced when fossil fuels are burned

fertilizers chemicals that farmers put on crops to make them grow

fossil fuel a fuel such as oil, coal, or natural gas that is made from the remains of prehistoric living things

landfill big holes in the ground where trash is buried

nutrients natural substances that help living things grow and thrive

pesticides chemicals that farmers put on crops to stop insects and other animals eating the plants

recycle to take waste materials and reuse them by turning their materials into new objects

sewage waste such as poop or dirty water from homes and factories

FIND OUT MORE

Books

Infographics: Pollution, Alexander Lowe, Cherry Lake Publishing, 2022

Pollution (Earth in Danger), Emily Kington, Hungry Tomato, 2022

Studying Pollution (Citizen Scientist), Izzi Howell, PowerKids Press, 2022

Websites

climatekids.nasa.gov/air-pollution/

kids.niehs.nih.gov/topics/pollution/index.htm

natgeokids.com/uk/kids-club/cool-kids/general-kids-club/tips-to-reduce-plastic-pollution/

INDEX

Answers: 1. a; 2. c; 3. b; 4. c